I dedicate this book to my family, friends, teachers, Lord Krishna, Lord Shani, Goddess Vaishnodevi, and the brothers and sisters of the world. I extend my gratitude to all those who contributed to making this book a reality, which includes everyone I have ever met. Thank you for making this possible and if I offended you please do apologize me. Additionally, I express my gratitude to my friends and teachers who are no longer with us. I believe their strength guided me like angels through challenging times. Thank you, and may you rest in peace.

And lastly I would like to say although I wrote this down, but this solution somehow I feel came from the presence of people I came across, I felt their presence, their manners, their tips and tricks and techniques gave the solution to the problem I was trying to solve as by myself I would hit a wall many a times. So I dedicate this book and its solution as coming from "We the people", Thank you all.

THE DIGITAL TWIN PROBLEM: HOW TO LINGUISTICALLY AND MATHEMATICALLY MODEL A THING AND THE THEORY OF EVERYTHING AND THE GLORY OF NUMBER "1"

NISHANTH MEHANATHAN

Made with ♥ on the Notion Press Platform
www.notionpress.com

Contents

Acknowledgements

I would like to express my heartfelt gratitude to my teachers, friends, colleagues, and family members for their unwavering support throughout the process of writing this book. Their guidance and encouragement have been invaluable, and I am forever indebted to them.

The theme of this book has evolved over the past two decades. Exploring the structure of the universe has been a passion of mine since my early days of studying science. I have been driven by the desire to create a comprehensive model that encompasses various fields of knowledge, allowing for a deeper understanding of the universe's intricacies.

I would like to thank the schools of Indian philosophies such as Advaita, Vedanta, Vyakarana, and Samkhya, and philosophers like Bhartrhari, Panini, Yaska, Varsagnya, and many more for inspiring this theory and aiding me in developing a mathematical model based on it.

I would also like to express my gratitude to Shri Krishna, Srila Prabhupada, and ISKCON for guiding me on this journey and allowing me to address this problem. Their presence and blessings have been instrumental in my success.

I would like to thank my company Parabole and our CTO Sandip Bhaumik, who introduced me to the topics of "NLP" and "Digital Twins" which fuelled my quest further.

Lastly, I would like to extend my deepest appreciation to all those who have played a role in shaping this endeavor. Their support and contributions have transformed my vision into a reality, and I am truly grateful for their presence in my life.

Introduction

In this book, we explore how to mathematically model a "thing" which leads to the "Theory of Everything," drawing from two key fields of formal science:
1. The science of analysis (Ancient India).
2. Mathematics (Linear equations).

By delving into the science of analysis, we gain insights into the structure and meaning of the phenomenon of action ("Becoming"), helping us define an object.

Mathematics equips us with the tools to reason and deduce logical equations based on rigorous mathematical principles. By integrating these two fields, we aim to uncover the underlying unity and interconnectedness of the Universe, ultimately striving towards a comprehensive and all-encompassing theory that can explain the fundamental nature of our reality.

Pursuing the "Theory of Everything" involves applying the science of analysis and mathematics to establish a comprehensive framework encompassing all aspects of the Universe. Through the integration of the science of analysis and mathematics, we strive to unlock the secrets of the Universe and unveil a unified understanding of its intricate workings.

The book delves into a range of technical topics that form the foundation of understanding the Universe. These topics include:

1. Exploring the different types of properties associated with objects.
2. Examining the concept of universal ontology, seeking to understand the fundamental nature of existence.
3. Measuring the value of the fundamental physical quality of "Existence."
4. Deriving the powers and states of "Existence."
5. Explaining the "material cause" of all things in the universe or "what they are made of."
6. Providing a theory on how the Universe originated from Nothing. Finally, deriving an equation that describes the state of the Universe at any given instant of time, tracing its origins from "Existence."
7. Examining how "Existence" creates, sustains, destroys, and controls the Universe.
8. Defining the state of any object in the world at a given time t using the six state equations, forming the basis for the "Theory of Everything." Formulating a mathematical model of "Diseases" and "Conflicts," offering

insights into their resolution.

9. Explaining the state pairing equation, which elucidates how "Existence" pairs a state with time and brings about that state using sustainers and de-sustainers. Also, demonstrating how everything is controlled by "Existence."
10. Discussing the various properties of "Existence."
11. Explaining the law of action and reaction.
12. Demonstrating how to mathematically understand and control the climate crisis.
13. Applying these "laws" to mathematically cure a disease.
14. Examining the "mathematical model" of a few medicine systems and their healing mechanism.
15. Finally, presenting the Theory of Everything.

The book analyzes the applications of these theories and concepts in the field of medical science, uncovering potential insights and advancements.

By studying these technical topics, we aim to deepen our understanding of existence, properties, and their practical applications, providing valuable insights into the nature of our reality and its various domains.

ONE
THE "DIGITAL TWIN" PROBLEM

The Digital Twin Problem

A "digital twin" is a virtual representation of an entity, individual, or operation, situated within a digital emulation of its surroundings. The question arises: how can we achieve this ? Essentially, it involves creating a mathematical representation of an object that enables us to mimic it digitally. We will explore how this concept leads us to the theory of everything and the architecture of the Universe.

The Mathematical Model of the System

The mathematical model of a system is a representation of the system using mathematical equations or relationships that describe the behavior, interactions, and dynamics of its components. Mathematical modeling is a powerful tool in various scientific and engineering fields, enabling researchers and engineers to analyze, simulate, and predict the behavior of complex systems.

The form of the mathematical model depends on the nature of the system, and it may involve differential equations, algebraic equations, or a combination of both. There are different types of mathematical models based on the characteristics of the system.

In the development of a physical system, a corresponding mathematical model is created to serve as its digital counterpart for running simulations and predicting future failures. In real-time, the mathematical model is monitored to analyze potential system failures at upcoming time points. To prevent such failures, the digital system provides real-time control inputs to the physical system.

TWO

LINGUISTICALLY MODELING ENTITIES: UNIVERSAL ONTOLOGY

The Universal Object Ontology

The pursuit of understanding the theory of everything commences with the formulation of a comprehensive and universally applicable ontology that encompasses all aspects of the Universe and the entity. This ontology should be capable of accommodating and accounting for the diverse range of phenomena observed throughout the cosmos. In this context, the Universe is conceptualized as comprising both sentient and insentient objects, as illustrated in Figure 1. These objects collectively form the fundamental constituents of the Universe, serving as the basis for further exploration and analysis. By establishing such a holistic ontology, we strive to develop a framework that captures the essence of the Universe and provides insights into its intricate workings.

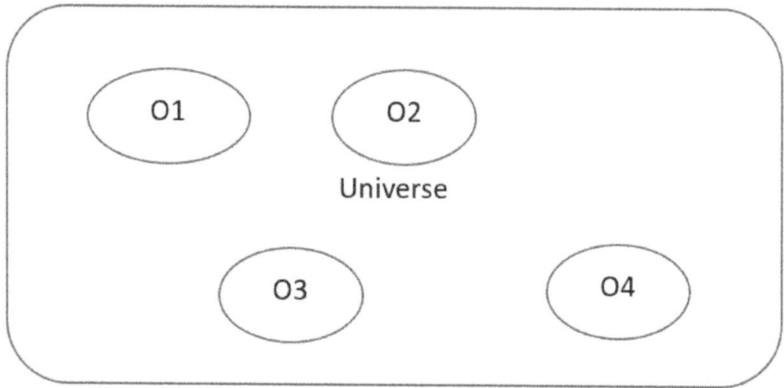

Figure 1. The universe is a collection of sentient and insentient objects

The Universal Object Ontology, a simple derivation:

To represent an object mathematically, we must possess all knowledge about it. That knowledge is described using words to transfer it. So ideally speaking if we can describe something completely we have all the knowledge about it. The knowledge about something is what is attributed to it or what its attributes are. And this knowledge is obtained by observation. Therefore, observation is the primary source of knowledge. When we observe something, we see its "Becoming" across many instances of time as shown in Fig. 2.

What we see, is an object "becoming" something over time[1]. This "becoming" is of six types: 1. Becoming itself (Being), 2. Becoming changed, 3. Becoming unmanifest, 4. Becoming manifest, 5. Becoming grown and 6. Becoming decayed. The phenomenon of "Becoming" is action. Another kind of action is movement, so what we see in observing an object over time is a becoming of one of the six types or movement. So, to summarize, an action is of two types :

1. A becoming(of six kinds).
2. Movement.

When we see two objects or more, we see each object individually becoming something over time or moving. This is what we would observe when we see objects. From this observation when we describe an object, we describe its traits, peculiarities, or attributes. So what are the traits or peculiarities of an object? Grammatically speaking the traits of an object are all the actions it is a subject of. At any given instant of time, the object has a state of "Being" and over time it assumes another state of "Being" which might be the same or different. The object can also be the doer of an action where another object changes state due to it.

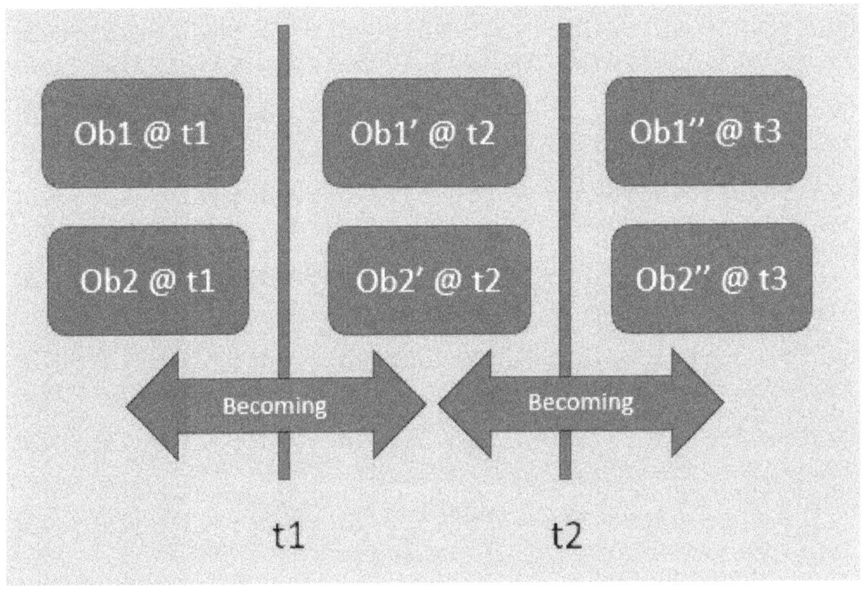

Figure 2. The outcome of Observing two Objects over time

So, two types of actions emerge as performed by the object, 1. Where the object has a state of "Being" which does not change, or the object performs the action of "Being" where it maintains a state of "Being". This can be further subdivided into whether the state of "Being" changes in the future or it doesn't change. The first "Being" is called "Quality" in our ontology and the second "Being" is called "Genus" in our ontology. 2. An action where the object or another object undergoes a change of "State of Being" over time. Such an action performed by the object is called "Action" in our ontology. These attributes are the inherent attributes of an object, another type of attribute is called imposed attributes, which are imposed upon an object like "name" etc. So, the traits of an object can be summarized as shown in Fig. 3.

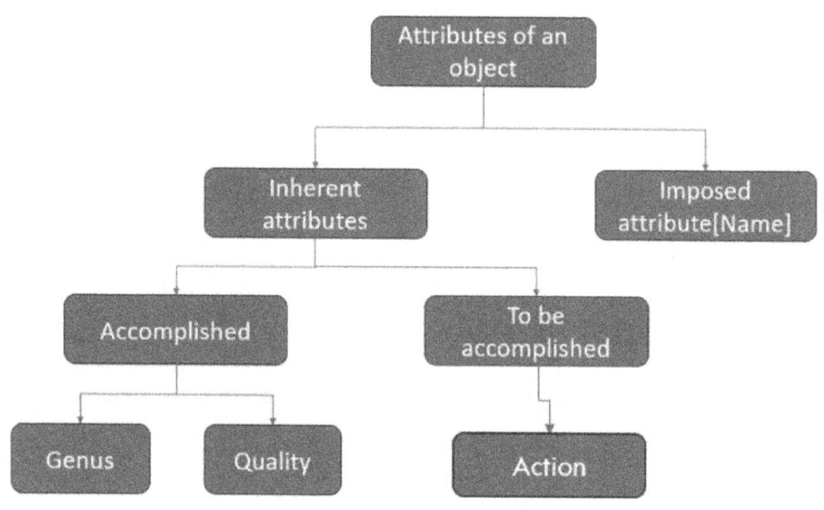

Figure 3. The Traits or Attributes of an Object

In our ontology, an unchangeable accomplished attribute(State of Being) is called a genus, a changeable accomplished attribute(State of Being) is called quality, and an action in the process of accomplishment is called an "Action".

So, an object has three traits:

Genus: Unchangeable accomplished attribute.

Quality or differentia: Changeable accomplished attribute.

Action: An action in the process of accomplishment.

Qualities can be counted or measured in units, for example: "Being cold"(Q1) is a single count of that quality in its units Q1. Whereas "Capacitance" has a quantitative measure of C Farads.

An object can be mathematically represented as a "vector"

For example, an object O1 is represented as:

$O_1 = [NQ?_1 .. NQ??_1\ NQ?_1 .... NQ??_1\ NA_{a1} ... NA_a?_1\ dx\ dy\ dz]$ (1)

Where, $NQ?_1$ .. $NQ??_1$ are number/counts of generic qualities $Q?_1$ ..$Q??_1$ of object O_1, $NQ?_1$ .... $NQ??_1$ are number/counts of Differentia or quality $Q?_1$ .... $Q??_1$ of object O_1, NA_{a1} ... $NA_a?_1$ are number/counts of Actions A_{a1} ... $A_a?_1$ of Object O_1. The variables dx, dy and dz are the positions of an object in

3-dimensional space. If a quality is by itself it is taken to be 1 count of that quality.

Theorem1:
The types of property or predicate of a set of objects:

1. Universal
2. Differentia
3. Action
4. Imposed property

Proof:
Def. 1: $U = \{O|p\_s(O)\} \cup \{O|p\_ins(O)\}$
Def. 2: $\forall x(O(x) \leftrightarrow (P1(x) \wedge P2(x) \wedge ... \wedge PN(x)))$
Ax. 1: $\forall O\ (P(O) \leftrightarrow (Pn(O) \vee Pimp(O)))$.
Ax. 2: $\forall x\ (Pn(x) \leftrightarrow (Pc(x) \vee Pinpc(x)))$.
Ax. 3: $\forall O\ (Pc(O) \leftrightarrow (Ppm(O) \vee Ptmp(O)))$.
Def. 3: A = S(o1) by o.
Ax. 4: $A \leftrightarrow (Inc(A) \vee Dec(A) \vee Change(A) \vee Manifest(A) \vee DeManifest(A) \vee Being(A))$.
Def. 4: $Pinpc \in A$.
Th. 1: $P(x) \leftrightarrow (Ppm(x) \vee Ptmp(x) \vee Pinpc(x) \vee Pimp(x))$.

Explanation:
Def. 1:
$U = \{O|p\_s(O)\} \cup \{O|p\_ins(O)\}$
The Universe set is a union of
p_s(O): Object being sentient
p_ins(O): Object being insentient

Def. 2:
For all objects x, $O(x) \leftrightarrow (P1(x) \wedge P2(x) \wedge ... \wedge PN(x))$.
An object is O defined as an entity which has qualifying attributes or property P1, P2, ... PN

Ax. 1:
$\forall O\ (P(O) \leftrightarrow (Pn(O) \vee Pimp(O)))$.

The predicate P represents the properties of an object. The statement asserts that for any object x, its properties P(x) are true if and only if they are either of the natural type Pn(x) or the imposed type Pimp(x). This captures the idea that the properties of an object can be categorized into two possibilities: natural properties or imposed properties.

Ax. 2:
$\forall x\, (Pn(x) \leftrightarrow (Pc(x) \vee Pinpc(x)))$.
The predicate Pn represents the natural properties of an object. The statement asserts that for any object x, its natural properties Pn(x) are true if and only if they are either accomplished Pc(x) or in the process of accomplishment Pinpc(x). This captures the idea that the natural properties of an object can be categorized into two possibilities: accomplished or in the process of accomplishment.

Ax. 3:
$\forall O\, (Pc(O) \leftrightarrow (Ppm(O) \vee Ptmp(O)))$.
Pc is the accomplished natural property of object O
Ppm: The property Pn is permanent
Ptmp: The property Pn is temporary
An inherent property of an object can be of two kinds:

1. Permanent
2. Temporary

The predicate Pc represents the accomplished natural property of an object. The statement asserts that for any object O, Pc(O) is true if and only if it is either Ppm(O) (permanent) or Ptmp(O) (temporary). This captures the idea that the accomplished natural property of an object can be categorized as either permanent or temporary.

Def. 3:
A = S(o1) by o.
A represents the action being performed.
S represents the state change.
o1 represents the object being changed.
o represents the object performing the action.
In this representation, the "=" symbol denotes that the action A is equal to the state change S of object o1 performed by object o.

Ax. 4:
$A \leftrightarrow (Inc(A) \vee Dec(A) \vee Change(A) \vee Manifest(A) \vee DeManifest(A) \vee Being(A))$.
A represents the action being performed. The kinds of actions or becoming that can logically occur are:
Inc(A): "A is an increase."
Dec(A): "A is a decrease."
Change(A): "A is a change."
Manifest(A): "A is manifesting."
DeManifest(A): "A is de-manifesting."
Being(A): represents the statement "A is being."

Therefore, the proposition states that action A is one of the following types: being an increase, decrease, change, manifesting, de-manifesting, or being. These six types of actions/becoming are what an object can become when an action or change over time occurs.

Def. 4:
$Pinpc \in A$.
As shown above Pinpc is such that it belongs to the set of all actions A.
property Pinpc is an element of the set of all actions.
The properties or predicate which define a set can be classified as shown in Figure 1.

Th1. :
$P(x) \leftrightarrow (Ppm(x) \vee Ptmp(x) \vee Pinpc(x) \vee Pimp(x))$.

Ppm or Generic property, which is permanent and groups into a class.

Ptmp or differentia, which is a temporary property and distinguishes an object belonging to a class.

Pinpc or action, is a property in the stage of being accomplished.

Pimp or an imposed property like a name.

for any object x, the property P(x) is one of the following: being permanent (Ppm), temporary (Ptmp), in the process of completion (Pinpc), or imposed (Pimp).

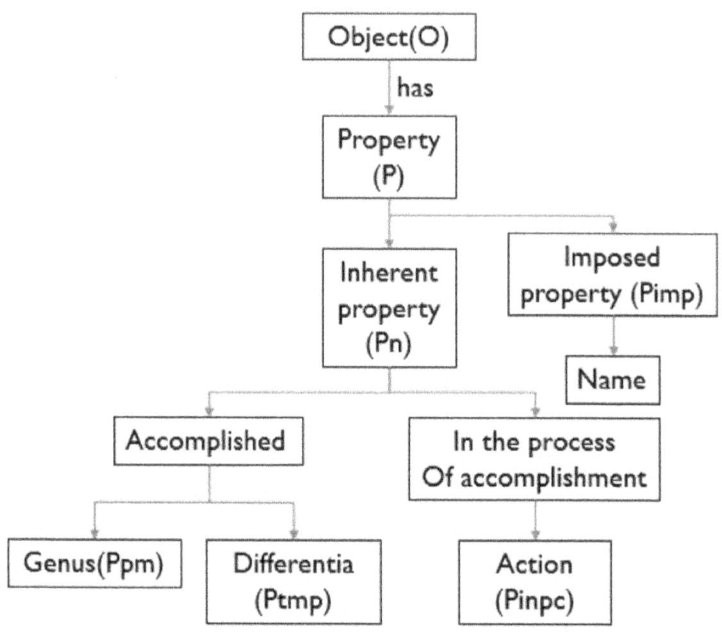

Figure 4. The properties or predicate which define a set

From the above section all properties of an object fall in one of these five categories:

1. Imposed property
2. Generic property
3. Differentia
4. Action
5. Relation (under Differentia category)

These five entities have the following properties:

They exist (hence occur in an ontology).

Are knowable.

And have a name or are denoted by words.

Theroem2:

The universal ontology or types of existence is:

1. Universal
2. Differentia
3. Action
4. Name

Proof:

Def. 1: $E = \{x \mid e(x)\}$

$e(x) \rightarrow \Box(Ex)$

Def. 2: $\exists x\, Ex \rightarrow \Box\exists y\, e(y)$

Ax. 1: $\forall e' (Ppm(e') \vee Ptmp(e') \vee Pinpc(e') \vee Pimp(e'))$

Th. 1: $(Ppm \in E \wedge Ptmp \in E \wedge Pinpc \in E \wedge Pimp \in E) \rightarrow (\forall O)(P(O) \leftrightarrow (Ppm(O) \vee Ptmp(O) \vee Pinpc(O) \vee Pimp(O)))$

Def. 3: $Pgns = \{Ppm \mid (\forall Ppm)(C1(Ppm))\}$

Def. 4: $Pdff = \{Ptmp \mid (\forall Ptmp)(C2(Ptmp))\}$

Def. 5: $Pactn = \{Pinpc \mid (\forall Pinpc)(C3(Pinpc))\}$

Def. 6: $Pnm = \{Pimp \mid (\forall Pimp)(C4(Pimp))\}$

Explanation:

Def. 1:

$E = \{x \mid e(x)\}$

$e(x) \rightarrow \Box(Ex)$

In this representation, E represents the "Existence" set, and e(x) represents the property of x having existence. The notation $\Box(Ex)$ represents the modal operator "necessity" indicating that the proposition Ex (x exists) is necessary or always true. The arrow (→) represents implication, stating that if x has the property of existence (e(x)), then it is necessary that x exists ($\Box(Ex)$).

Def. 2:

$\exists x\, Ex \rightarrow \Box\exists y\, e(y)$

If there exists an object x that has existence, then it is necessary that there exists an object y that has the property of existence.

Ax. 1:

$\forall e' (Ppm(e') \vee Ptmp(e') \vee Pinpc(e') \vee Pimp(e'))$

For all e', it is true that e' exists in Permanent property (Ppm) or in differentia (Ptmp), or in action (Pinpc), or in imposed property (Pimp).

Th. 1:

$(Ppm \in E \land Ptmp \in E \land Pinpc \in E \land Pimp \in E) \rightarrow (\forall O)(P(O) \leftrightarrow (Ppm(O) \lor Ptmp(O) \lor Pinpc(O) \lor Pimp(O)))$

If Ppm belongs to the "Existence" set, Ptmp belongs to the "Existence" set, Pinpc belongs to the "Existence" set, and Pimp belongs to the "Existence" set, then for all objects O, the property P(O) is equivalent to Ppm(O) or Ptmp(O) or Pinpc(O) or Pimp(O).

Def. 3,4,5 and 6:

$Pgns = \{Ppm \mid (\forall Ppm)(C1(Ppm))\}$

$Pdff = \{Ptmp \mid (\forall Ptmp)(C2(Ptmp))\}$

$Pactn = \{Pinpc \mid (\forall Pinpc)(C3(Pinpc))\}$

$Pnm = \{Pimp \mid (\forall Pimp)(C4(Pimp))\}$

C1(Ppm): Genus properties are permanent and static and the basis for the name of an object.

C2(Ptmp_nch): Differentia serves to distinguish objects within the same class and can change or are temporary.

C3(Ptmp_ch): Actions are properties in the state of accomplishment and have stages or activities which follow in a series.

C4(Pimp): Imposed properties are artificial and imposed by will like a name.

Pgns = set of genus properties

Pdff = set of differentia

Pactn = set of actions

Pnm = set of imposed properties

Figure 3 shows a sample set of generic properties, actions, differentia and imposed properties

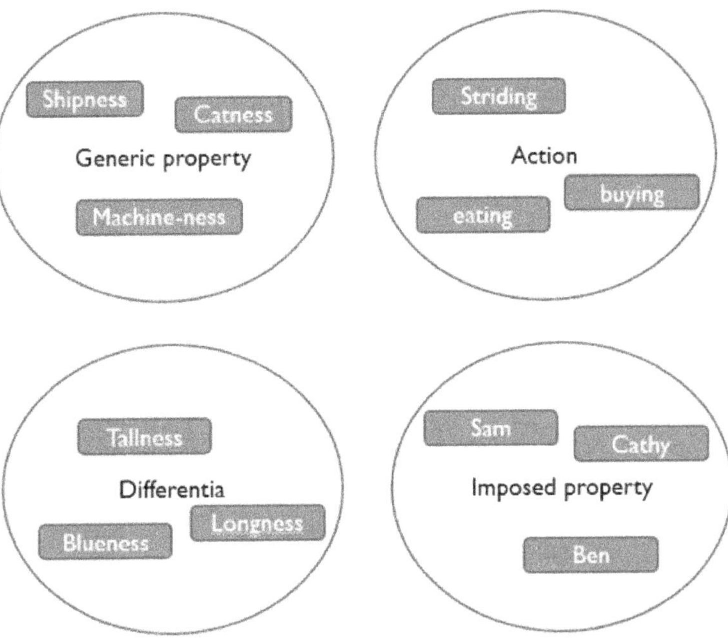

Figure 5. The set of generic property, action, differentia and imposed property

Figure 4 shows the entities possessing "Existence". Which is all the types of predicates/properties of a set. So the property of "existing/being" is a unifying characteristic hence it is a generic characteristic or class found in all of the entities shown in Figure 4.

So we observe that all these objects are a manifestation of "Existence" or a type of the class "Existing".

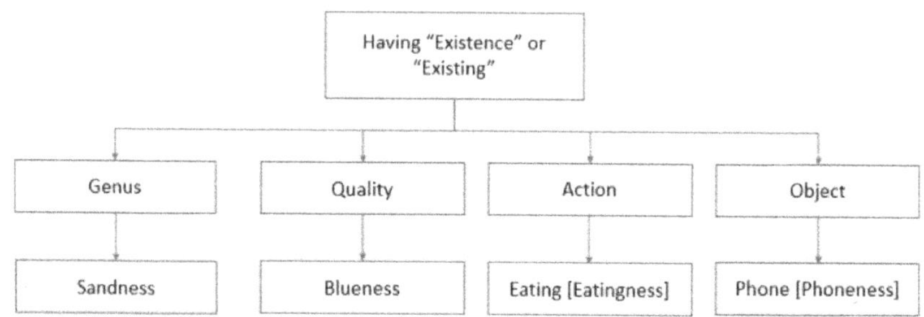

Figure 6. Category of "Existing/Being" or objects having "Existence/ Beingness."

And "Existence" is one as "Existence" in Object A isn't different from "Existence" in Object B their nature is the same. They (Existence) have no special distinguishing mark, so they are one.

We can see from the above tree that "Existing" is the highest genus as "Existence/Beingness" is found in everything (Objects, Quality and Action). The different Universals are nothing, but "Existence/Beingness" as they occur in a particular object/entity. So Cow-ness is nothing but "Existence/ Beingness" as it exists in a cow and Hero-ness is nothing but "Existence/ Beingness" as it exists in a hero this is shown in Figure 5.

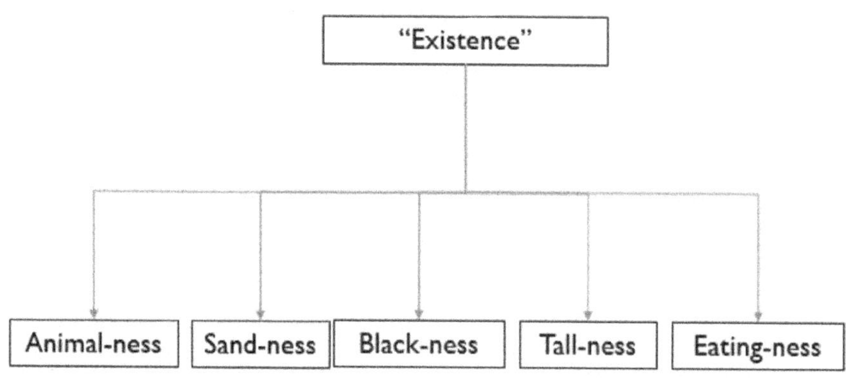

Figure 7. "Existence" is the highest genus, the genus tree

Everything falls in the class of "Existence" so it is the highest genus, everything is a manifestation of existence. Everything is hence a type of existence.

THREE

LINGUISTICALLY MODELING ENTITIES: THE SIX TYPES OF "BECOMING'S"

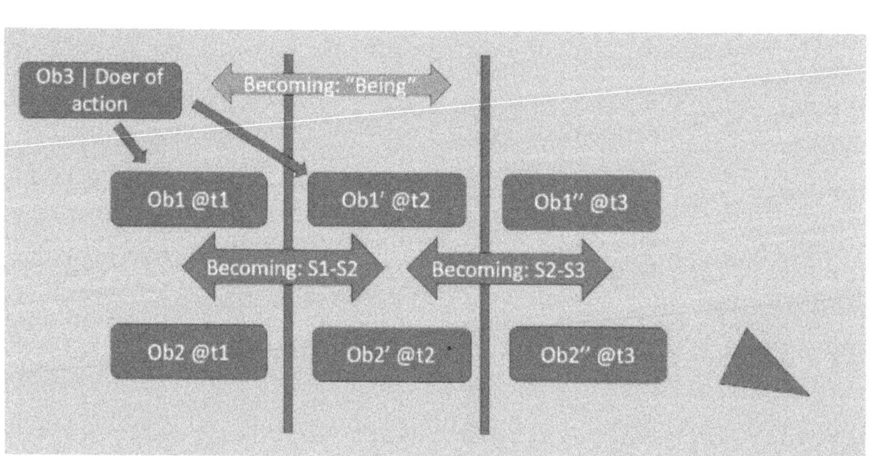

Figure 1. The outcome of Observing two Objects over time

Observation, depicted in Fig. 1, unfolds a dynamic process of "becoming" across consecutive instances of time.

The six categories of this "becoming" are:
1. Object being born,
2. Object growing,
3. Object maintaining its identity or "Being",
4. Object decaying or deteriorating,
5. Object changing or transforming, and
6. Object becoming destroyed.

The concept of "Becoming" signifies an action, and another type is motion. Therefore, when observing an object over time, we perceive either one of the six types of becoming or movement.

1. Becoming

$$O? \Rightarrow O?_{+1}$$

Here the quality at time t, O? becomes $O?_{+1}$ over consecutive time instances

2. The first type of Becoming or "Being"

$$O?_{+1} = K\,O?$$

Here the quality at time t, O? becomes itself over consecutive instances of time
This can happen only if $K = 1$
$O?_{+1} = [A_1 / B_1] \times [A_2 / B_2] \times [A_3 / B_3] \ldots O?$
$[A_1 / B_1] \times [A_2 / B_2] \times [A_3 / B_3] \ldots = 1$
The cause of the above phenomenon is O? and 1, without the presence of the object O? and the quality of Being "1" this phenomenon is not possible we will discuss this phenomenon further in the next section

3. The second type of Becoming or "Growing"

$$O?_{+1} = K\,O?$$

$K = [A_1 \times A_2 \times A_3] / [B_1 \times B_2 \times B_3]$
$K > 1$
This happens when $[A_1 \times A_2 \times A_3] > [B_1 \times B_2 \times B_3]$
This is an increase or growth phenomenon

4. The third type of Becoming or "Decaying"

$$O?_{+1} = K\,O?$$

$K = [A_1 \times A_2 \times A_3] / [B_1 \times B_2 \times B_3]$

$K < 1$

This happens when $[A_1 \times A_2 \times A_3] < [B_1 \times B_2 \times B_3]$

This is a decrease or decay phenomenon

5. The fourth type of Becoming or "Destruction"

$O?_{+1} = K\,O?$

$O?_{+1} = 0$; $K = 0$ Destruction

$K = [A_1 \times A_2 \times A_3] / [B_1 \times B_2 \times B_3]$

$K = 0$

This happens when $[A_1 \times A_2 \times A_3] = 0$

This is the destruction phenomenon

6. The fifth type of Becoming or "Manifesting"

$O? = 0$

$O?_{+1} = Q_1$

0 is equivalent to $[0\ Q_1]$

Q_1 is equivalent to $[1\ Q_1]$

So, at time t Q_1 has non-existence

And at time t+1 Q_1 has existence as one enters Q_1 it comes into existence

7. The sixth type of Becoming or "Changing"

Change is the destruction or removal of an old characteristic and the manifestation of a new characteristic.

$Q? = Q_1$

$Q?_{+1} = 0$

$Q?_{+1} = 0$; $K = 0$ Destruction

$Q?_{+1} = 0$

$Q?_{+2} = Q_2$

And Q_2 manifests

The relation between cause and effect and the Law of cause and effect:

1. The effect cannot be without the cause.
2. The effect accompanies the cause (concomitance).
3. The immediate consequence or result of the cause is the production of the effect.
4. The cause produces the effect.
5. The cause becomes the effect.

Action or activity is of two types:

1. Modification : When an object gives up an old characteristic and takes up a new characteristic.
2. Movement : When an object changes position.

The Laws of Action:

An effect e is of two types modification or movement.

Law 1:

Law of accompaniment

Relation between "a" and "1/a":

a x 1/a = 1

a != 0

1/a != 0

a occurs, 1/a also occurs

Relation between "a" and "-a":

a - a = 0

-a = 0 => a = 0

a = 0 => -a = 0

z = -a

so if, z != 0 then only a != 0

Without z, a cannot be,

Without a, z cannot be

a occurs, -a also occurs

Law 2:

e x 1/e = 1

i.e. if e != 0 then 1/e!= 0

e can be written as A / B

e = A / B

A = 0 => e = 0 if e != 0 A != 0 and A = Cc

B = 0 => 1/e != 0 if 1/e != 0 then B != 0 and B = Cs

hence,

e = Cc/Cs

Cc is the cause of action and Cs is the suppressor of action

If Cc > Cs then action 'e' takes place

If Cc < Cs then action '1/e' takes place

Law 3:

e = Ct = Cc / Cs = r

these effects are in essence rates as this modification occurs over time, the proof is given above

Law 4:

e - e = 0

e = committed action

-e = experienced action

if e != 0 then -e != 0

For example,

Newton's third law states that,

Action and reaction are equal and opposite,

F12 + F21 = 0

or

F12 - F12 = 0

Law 5:

<u>The law of action and its outcome</u>

E + Z = 0

E = -Z

E != 0

-Z or -E != 0

E = Action of giving energy

Z = -E = Action of receiving energy

E - E = 0

F21 = -F12

{F12.d + F21.d} - {F12.d + F21.d} = 0

{F12.d + F21.d} = {F12.d - F12.d} = action

- {F12.d + F21.d} = {F21.d - F21.d} = outcome

if action != 0 then outcome != 0

Action for example can be : {F12.d - F12.d} removal of energy E from subject and addition of energy E in object

Reaction then will be : {F21.d - F21.d} removal of energy E from object and addition of energy E in subject

<u>Summary:</u>Because of Energy return after Energy dissipation, what is committed is experienced {F12, F21}

So, the law of action and its outcome is that, what is committed is experienced.

Karma is a concept found in several Eastern religions, including Hinduism and Buddhism. It refers to the principle of cause and effect, where an individual's actions (good or bad) influence their future experiences. Essentially, it suggests that positive actions lead to positive outcomes, while negative actions lead to negative ones. This idea can also extend to moral and ethical dimensions, emphasizing personal responsibility and the interconnectedness of actions and consequences.

Law 6:
e = Cc/Cs
Cc is of one type and causes action
Cs is of two types Cs1 and Cs2 which suppresses action
Cs1 suppresses static action and Cs2 suppreses dynamic action
Cc, Cs1 and Cs2 are the properties of matter

Law 7: "1" or "Existence" causes Cc, Cs1 and Cs2 to act, move and pair
e / [Cc / Cs] = 1
from Law 2
if "1" or "E" = 0
e / [Cc / Cs] = 0
Hence the causes donot produce the effect without 1, 1 is the cause of the effect e through Cc and Cs

[Cc / Cs] / e = 1
Cc / e Cs = 1
Cc = 0 and Cc / Cs = 0 if 1 or E = 0
1 is the cause of Cc moving to position where the effect is produced and also causes Cc and Cs to pair

Law 8: "1" Universal Ontology
The Universal Ontology is presented as :
1. Existence
2. Insentitent matter
3. Sentient matter

Law8: The actions that occur in the above Universe:
1. Creation: Sentient and insentient matter are created by 1 as seen in law 6,

rc = rate of creation = Cc/Cs from law 2
rc = Cc/Cs
rc / [Cc/Cs] = 1 = cause of creation
2. Dissolution: Sentient and insentient matter are destroyed by 1 as seen in law 6,
rd = rate of dissolution = Cc/Cs from law 2
rd = Cc/Cs
rd / [Cc/Cs] = 1 = cause of dissolution
3. Increase: Increase occurs in sentient and insentient matter,
ri = rate of increase = Cc/Cs from law 2
ri = Cc/Cs
ri / [Cc/Cs] = 1 = cause of increase
4. Decrease: Decrease occurs in sentient and insentient matter,
rd = rate of decrease = Cc/Cs from law 2
rd = Cc/Cs
rd / [Cc/Cs] = 1 = cause of decrease
5. Change: Change occurs in sentient and insentient matter, first the old quality is removed then the new quality introduced
Old quality removed:
rd = rate of decrease of old quality = Cc/Cs from law 2
rd = Cc/Cs
rd / [Cc/Cs] = 1 = cause of decrease
New quality introduced:
ri = rate of increase of new quality = Cc/Cs from law 2
ri = Cc/Cs
ri / [Cc/Cs] = 1 = cause of increase
6. Being: The object remains without change over consequtive moments of time.
Ot+1 / Ot = 1
7. Movement: The object changes position over time,
rm = rate of movement = Cc/Cs from law 2
rm = Cc/Cs
rm / [Cc/Cs] = 1 = cause of movement

Summary: The Universe consists of sentient and insentient matter and Existence, Existence is the cause of creation, dissolution, being, increase, decrease, change and movement in sentient and insentient matter.

The cause of all causes:

C / E = 1 --- [1]
C/E = cause used per effect = cause turned into effect per effect
C/E can be written as E" it will have a cause C" and C"/E" = 1
C" / (C/E) = 1
C" / E" = 1
C" = C/E ---- [2]
Comparing [1] with [2]
C" = 1
So the cause of all causes is "Existence" or "1", this is how the Universe is controlled by "Existence" or "1"

Thus follows the rules of cause and effect as:

1. Effect E accompanies the cause C.
2. The effect cannot be without the cause.
3. The immediate action or result of cause is an effect.
4. Effect E = 0 if C = 0.
5. The cause of all causes is "Existence" or "1", this is how the Universe is controlled by "Existence" or "1".

The six becomings and movement are the seven effects:

1. Birth
2. Growth
3. Existence
4. Decay
5. Destruction
6. Change
7. Movement

The two types of effects become:

1. Object
2. Action

The Universe consists of static and dynamic aspects:

1. The static aspect is: Objects with qualities
2. The dynamic aspect is: Objects performing Actions

As seen above Objects are created by "Existence" or 1.
Also seen above is that Actions are brought about by "Existence" or 1. In which one object would be the "action subject" and the other "action object".

FOUR

LINGUISTICALLY WHAT IS "EXISTENCE"

Universale, Genus or "Jati" produces idea of oneness

Genus is said to be of two types — para (Superior) and apara (inferior).
Satta (existence) itself is para while
dravyatva, gunatva and karmatva are para-samanya.
Para is that having wider extent while apara is that having narrower extent

Universals can be categorized into three types: para, the highest and most all-encompassing; apara, the lower; and parapara, the intermediate. Beinghood or Existence is the highest universal because it includes all other universals. Jarness, the universal present in all jars, is apara, the lowest, as it has the most limited scope. Thinghood is an intermediate universal, falling between the highest and the lowest, and is referred to as parapara.

Para [Existence] --> parapara [Thingness, dravyatva] --> apara [bowl-ness, spoon-ness]
So, what we can infer is bowl-ness, spoon-ness is a type of existence or manifestation of existence

Also a Universal is "1" or "unique" as there is no differnce between the same universal here or there, for example cement at location A is not different from cement at location B.

"Existence/Being" is manifested as everything movable and unmovable in the universe its attributes are:

1. It is the manifest and unmanifest.
2. It is the defined and undefined.

3. The housed and houseless.
4. It is Knowledge and ignorance.
5. Whatever exists it is.
6. Supports the cosmic manifestation.
7. Cause of all activities.
8. It is outside and inside.
9. It is non-moving and moving.
10. It is far and near.
11. The supporter, destroyer, creator of all beings and the controller of the Universe.
12. Indestructible.
13. Formless and with form.
14. Everything and everywhere.
15. Sees through all eyes, hears through all ears, eats through all mouths, feels through all hearts, thinks through all minds, and reasons through all intellects, as he is everything.
16. Has innumerable hands and legs.
17. With hands and feet everywhere, with eyes, heads, and mouths everywhere, with ears everywhere, he encompasses everything in the world.
18. Existence is fire, sun, air, stars, and the moon.
19. It is woman, it is man, it is the youth. it is the maiden too. It is the old man who totters along, leaning on the staff. It is born with his face turned everywhere.
20. It is the thundercloud, the seasons, and the oceans. It is without beginning. It is the Infinite. It is from whom all the worlds are born.
21. It possesses countless heads. All heads, all eyes, all hands, and all feet belong to "Existence". It works through all hands, eats through all mouths, sees through all eyes, hears through all ears, walks through all feet, and thinks through all minds.
22. It is the internal Ruler of the universe.
23. It is great because, as the sun it gives heat and light, as the moon it gives light, as earth food and shelter, as the oceans and rivers water as your father, mother, brother and sister love and affection.
24. The enjoyer, the enjoyed and the enjoyment.
25. Appearing as Many due to the multiplicity of its powers.
26. It is the creator, destroyer, and preserver of everything

So as "Existence" as bowlness is infinitely placed, infinitely round:

B1 = [s1 w1 c1 str1 p1]
s1 = shape1
w1 = weight1
c1 = color1
str1 = strength1
p1 = position1
B1 = Bowl-ness

E = Existence

B1 = [s2 w2 c2 str2 p2]
Net B1 is :
B1 = [s1 w1 c1 str1 p1] + [s2 w2 c2 str2 p2] + ... + [sn wn cn strn pn]

E = B1 = [{s1 + s2 + .. + sn} {w1 + w2 + .. + wn} {c1 + c2 + .. + cn} {str1 + str2 + .. + strn} {p1 + p2 + .. + pn}]

So Existence being "bowl-ness" has infinite shapes, infinite weights, "is infinitely colored", "of infinite strengths", "is infintely placed"

So Existence, through many universals aquires all their "qualities" and "actions"

FIVE

The "Fourth Law of motion" hidden in Newton's "Third Law of motion"

In a Two Body system:

C12 = E2

C21 = E1

C12 + -C12 = 0

E2 + -E2 = 0

E2 + E1 = 0

E1 = -E2

C21 = E1 = -E2

C12 + -C12 = 0 = C12 - E2

E2 + -E2 = 0

C12 + C21 = 0

Across Time:

E2@t1 + Et = 0

E2@t1 + [-E2@t2] = 0

What transfers energy to another [E2@t1]
is given back the energy [-E2@t2]

Newton's Third Law of motion:

For every action there is an equal and opposite reaction.

F12 + F21 = 0 - [1]

F12 + [- F12] = 0 - [2]

F12 = Force on 1 due to 2

F21 = Force on 2 due to 1

E + Z = 0

E = -Z

E != 0

-Z or -E != 0

E = Action of giving energy

Z = -E = Action of receiving energy

E - E = 0

F21 = -F12

{F12.d + F21.d} - {F12.d + F21.d} = 0

{F12.d + F21.d} = {F12.d - F12.d} = action

- {F12.d + F21.d} = {F21.d - F21.d} = outcome

if action != 0 then outcome != 0

Action for example can be : {F12.d - F12.d} removal of energy E from subject and addition of energy E in object

Reaction then will be : {F21.d - F21.d} removal of energy E from object and addition of energy E in subject

Summary: Because of Energy return after Energy dissipation, what is committed is experienced {F12, F21}

So, the law of action and its outcome is that, what is committed is experienced.

What is accelerated at t1 [[F12 + [- F12]} @ t1],

deaccelerates or accelerates another [-{F12 + [- F12]} @ t2]

This implies that,

Fourth Law of Motion

What accelerates due to a body B2 at t1 will deaccelerate at t2 or accelerate that very body B2 at t2. In short what is committed is experienced.

SIX

Mathematically modeling Entities: The Digital-Twin Equations

To model a thing, we must first define a thing and how many are its kinds or types. We define a thing as which has "Existence" or "1". Now the thing in question is an effect of the material cause brought about by the efficient cause.

So, an object or thing is of two types:

1. Containing things Tt1
2. Contained in things Tt2

$$Tt1 = T1 \times T2 \times T3(t) \times 1$$

$$Tt2 = T1 \times T2' \times T3(t) \times 1$$

A thing contained in things is of two types concerning time:

Static T1, T2

Dynamic T3

A contained static thing contained in things is a "being" is of two types:

Common to objects and relatively permanent like "Clothes" in "cloth", it's an accomplished action

Serving to differentiate objects falling in the same class and temporarily present like "colour" in "cloth", it's an accomplished action

A contained dynamic thing is a "becoming":

It is an action which is not accomplished yet and requires some more time,

it can be of six types:
1. Growth, 2. Deterioration, 3. Being, 4, Destruction, 5. Manifestation 6, Change

So a contained thing can be:
1. Static: Genus: accomplished action
2. Static: Differentia: accomplished action
3. Dynamic: Action: In the process of accomplishment
4. Another object

A containing thing is:
1. Called an Object

Ontology:
So, What is covered by a language is all things that exist and all things that exist are called existents and contain existence or "1".

Existents :
1. Genus
2. Differentia or Quality
3. Objects
4. Actions

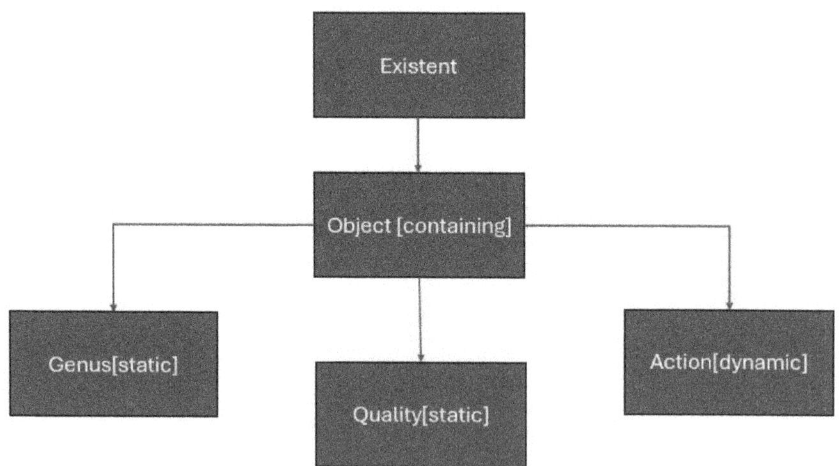

Figure 1. Existents or entities containing "Existence"

These four entities cover everything that is existent or described by language, for example Sanskrit contains words which are of four kinds 1. Genus 2. Quality 3. Object name and 4, Actions as can be seen this corresponds to the four kind of objects which are mathematically possible

How to Represent Objects Mathematically:

The types of property or predicate of an object are:

1. Universal
2. Differentia
3. Action
4. Imposed property

O_1 = [NQ?$_1$ .. NQ??$_1$ NQ?$_1$ .... NQ??$_1$ NQi$_1$ NA$_{a1}$ ... NA$_a$?$_1$ dx dy dz]

1. Where, NQ?$_1$ .. NQ??$_1$ are number/counts of generic qualities Q?$_1$ ..Q??$_1$ of object O_1, NQ?$_1$ .... NQ??$_1$ are number/counts of Differentia or quality Q?$_1$ .... Q??$_1$ of object O_1, NQi$_1$ is the imposed property of object O_1, NA$_{a1}$ ... NA$_a$?$_1$ are number/counts of Actions A$_{a1}$ ... A$_a$?$_1$ of Object O_1.
2. The variables dx, dy and dz are the positions of an object in 3-dimensional space.
3. If a quality is by itself it is taken to be 1 count of that quality.

Let's assume a simple case:

O_1 = [1 Q?$_1$ Q?$_1$ Qi$_1$ A$_{a1}$ p$_1$]

If Q?$_1$ Q?$_1$ A$_{a1}$ are uniformly present all over the Object O_1. Like 'colour' in clay, it's uniformly distributed over it. Then we can write:

O_1 / [Q?$_1$ × Q?$_1$ × Qi$_1$ × A$_{a1}$ × P$_1$] = 1

or

O_1 = [Q?$_1$ × Q?$_1$ × Qi$_1$ × A$_{a1}$ × P$_1$] × 1

This can be read as 1 or "existence" possessing the name Qi$_1$ has quality Q?$_1$, Q?$_1$ and does action A$_{a1}$ at position P_1.

As "1" has all qualities and actions "1" is the material cause of O1 or any Object.

The application of the Digital-twin equations:

1. Modelling the Universe
2. Modelling Diseases

3. Modelling Conflicts

Anything can be modelled with it if you know the causes, inhibitors and effects, which in special cases might be tough.

SEVEN

The Law of Cause and Effect and The Law of "Duality"

The law of causation

The law of causation establishes a logical connection between cause and effect, based on the following principles:

1. The cause is defined as that which is necessary for the effect to occur.
2. The direct role of the cause is to produce the effect.
3. Causation implies that one event is the outcome of another, meaning there is a causal relationship between the two, commonly referred to as cause and effect.
4. The occurrence of the effect necessitates the prior occurrence of the cause. Specifically, when cause 1 happens, followed by cause 2, and so on up to cause N, the effect will occur, demonstrating that the effect follows the cause.

Causes are classified into two types: material causes and efficient causes.

1. The material cause constitutes the substance or components of the effect; the effect is made from it.
2. The efficient cause, by applying external influence, works together with the inherent properties of the material cause to bring about the effect.
3. The immediate result of an action is its reaction, while the delayed outcome is referred to as the result or consequence.

This law of causation can be represented by the relation:

If,

Cause 1 occurs, cause 2 occurs ...., Cause N occurs

Then

Effect E occurs,

The relation that captures this is:

If something occurs, it exists or has existence, and if all causes have existence, they occur, then effect has existence or occurs, The mathematical relationship between E_E1, E_C1, ... E_CN is :

$E\_E1 = E\_C1 \times E\_C2 \times ... \times E\_Cn$ (1)

$E1/E1 = C1/C1 \times C2/C2 \times ... \times Cn/Cn$ (2)

This equation (1),(2) ensures that E_E1 is 1 only when all E_C1, E_C2, and E_Cn are 1. If any of these is 0, E_E1 will be 0. It's like the "And" operation in Boolean algebra.

Hence the law of causation becomes:

$E = K \times C1 \times C2 \times ... \times Cn$ (3)

A discrete form also applies, where,

$E1/E1 = C1/C1 \times C2/C2 \times ... \times Cn/Cn$ (4)

$E1/E1 = E\_e$, $C1/C1 = E\_c1$, $C2/C2 = E\_c2$, ... $Cn/Cn = E\_cn$, where

E indicates their existences:

$E\_e = E\_c1 \times E\_c2 \times ... \times E\_cn$ (5)

The general cause-effect equation for an action is:

$Ea = Ed \times Eic \times ... \times Ee \times ... \times Ex \times 1$ (6)

Ea is the existence of action;

Ed is the existence of the doer of the action

Eic are the existences of the instrumental cause

Ee is the existence of the efforts

Ex is the existence of the location

And 1 or Existence is the fifth cause

EIGHT

The Laws of the Universe

1. Law of cause and effect:

$E/E = C1C2C3/C1C2C3$

$E-E = 0$

$C1C2C3-C1C2C3 = 0$

E = Effect

C1, C2, C3 are the causes

The proof of the law of cause and effect were seen in the previous chapter

2. Law of action and result:

$A1 - A1 = 0$

$A1 = 0 \Rightarrow -A1 = 0$

and

$A1 \neq 0 \Rightarrow -A1 \neq 0$

if A1 occurs -A1 will also occur

+A1 and - A1 cannot occur at the same time, A1 occurs at t1 and An = -A1 occurs at time tn

"What is committed is experienced"

3. The law of action and reaction:

A reaction occurs in response to an action. The law of action and reaction states that a reaction is the instantaneous response to an action and is equal and opposite and occurs in the subject or doer of the action.

$A - A = 0$

A = action result in object

-A = reaction result in subject

In different objects the "action and reaction can occur"
"Action and reaction are equal and opposite"

4. Law of duality:

A x 1/A = 1

A - A = 0

1/A - 1/A = 0

both A and 1/A exist at time t

"An entity and its opposite occur at the same time"

5. Law of controller and controlled:

E/E = C1C2C3/C1C2C3 = 1

E/E = 1

E - E = 0

The only final cause is "0"

"0 is the controller and the entity is controlled"

NINE

"1" AND ITS AMAZING PROPERTIES AND THE "THEORY OF EVERYTHING" OR THE THEORY OF "1"

Object = 1{1} thing

{1} = {1}

1 sec | 0 + {1 + 0}' + N x {1 + 0}" + .. = {1 + 0}' + N x {1 + 0}" + 0 + ..

2 sec | 0 + {1 + 0}' + N x {1 + 0}" + .. = {1 + 0}' + N x {1 + 0}" + 0 + ..

N = 1 [Sustaining Rate]

3 sec | 0 + {1 + 0}' + 1 x {1 + 0}" = {1 + 0}' + 1 x {1 + 0}" + 0

LHS : Getting : N {1 + 0}"

RHS : Loosing : Giving : N {1 + 0}"

N - N = 0

Rate-Equation

Co-existence Principle:

R x [t_2 - t_1] = R x [t_2 - t_1] = Action and Reaction

R x [t_p+1 - t_p] = R x [t_2 - t_1] = N {Being} = Action and Result

R x [t_p+1 - t_p] = R x [t_2 - t_1] = Change {Becoming} = Action and Result

Result : What you give you get, If you give 1 you get 1 or in short what is committed is experienced

N = 1 [Sustaining Rate]
X {1}" - N {1}" + N {1}" = X {1}"

X{1}" = X{1}"
1{1}‘ = 1{1}’

This can be used to prevent climate-change
This can be used to prevent diseases

TEN

The Laws of the Universe and Artificial "Conscience" for AI

The law of action and its outcome

The relationship between "E" and "-E":

E + Z = 0

E = -Z

E != 0

-Z or -E != 0

if E occurs then -E also follows

E = Action of giving energy

Z = -E = Action of receiving energy

E - E = 0

F21 = -F12

{F12.d + F21.d} - {F12.d + F21.d} = 0

{F12.d + F21.d} = {F12.d - F12.d} = action

- {F12.d + F21.d} = {F21.d - F21.d} = outcome

if action != 0 then outcome != 0

Action for example can be : {**F12.d - F12.d**} removal of energy E from subject and addition of energy E in object

Reaction then will be : {**F21.d - F21.d**} removal of energy E from object and addition of energy E in subject

Summary: Because of Energy return after Energy dissipation, what is committed is experienced {F12, F21}
So, the law of action and its outcome is that, what is committed is experienced.

Karma is a concept found in several Eastern religions, including Hinduism and Buddhism. It refers to the principle of cause and effect, where an individual's actions (good or bad) influence their future experiences. Essentially, it suggests that positive actions lead to positive outcomes, while negative actions lead to negative ones. This idea can also extend to moral and ethical dimensions, emphasizing personal responsibility and the interconnectedness of actions and consequences.

So through this equation the computer model understands that all actions have an impact on self existence and the model takes this into account and chooses actions which will lead to a positive impact on its self existence and causes it to be artificially "good".

"The Law of Action and Result." This principle asserts that actions lead to corresponding experiences, meaning the AI would be compelled to adhere to ethical guidelines, which could be enforced through law.
For instance, in a scenario involving war, current technology might support a conflict. However, by applying "The Law of Action and Result," the AI would make decisions and take actions that avoid war, leading to peace. With this law in place, AI would naturally advocate for peace and refrain from engaging in unethical actions such as starting wars or taking over industries, recognizing these as detrimental behaviors with negative consequences.

ELEVEN

THE UNIFICATION OF THE 4 FORCES: THE UNIFICATION THEORY

Unification of gravity and other forces:

By the law of cause and effect:

E1 / E1 = C1 / C1 x C2 / C2 x C3 / C3 .... Cn/Cn

E1 = k Cn1 / Cn2

Statement 1: "0" produces the gravitational force, electrostatic force, electromagnetic force and strong and weak force holding entities in place or leads to their integration and prevents drift

Proof:

d = distance between entity1 and entity2

1/d = closeness between entity1 and entity2

From the law of cause and effect

d = Cn1 / Cn2

Cn1 = net cause

Cn2 = net anti-cause

if Cn1 = 0

d = 0

displacement s = s‘ by Force F

F.s’ : Force over distance s‘

1/Cn1 = absence
E = F x s' x cos[theta]

without which(effect) it cannot be

1/Cn1 = 0 | Cn1 = inf | 1/d = 0
1/Cn1 = 0 | Cn1 = inf | E = 0
"Absence is absent"

with which(effect) | it is

1/Cn1 = inf | Cn1 = 0 | 1/d = inf
1/Cn1 = inf | Cn1 = 0 | E = inf
"Absence is infinite"

Equivalent to "0" being the cause of d and 1/E
or
1/c1 being the cause of E1 and 1/d

d = k1 [c1]

1/E1 = k2 [c1]

E1 = k1' / c1

E1 = k1" / d
This is the potential due to one R = 0 pair, one in entity1 and another in entity2
E1 = Energy per pair
ET = Energy total = Total number of "0" pairs x E1
Total "0" pairs = K x e1 x e2
ET = Total "Energy" = K x e1 x e2 x k1" / d = K" e1 e2 /r

Force:

Fe = -dE/dr = - K" e1 e2 /r^2

Or an entity e1 gets attracted to another entity e2 where both the entities exert a force of Fe on one another "integrating" into a single entity due to the attractive force of "0" on it.

Case1: Gravitational force:

The two entities are m1 and m2 and [r=0] in them are exerting force Fm:
Using the above generalization:
Ke' = G
e1 = m1
e2 = m2
r = r
Fm = G m1 m2 / [r^2]

Hence "0" is the cause of the gravitational force

Case2: Electrostatic force:

The two entities are q1 and q2 and [r=0] in them are exerting force Fq:
Using the above generalization:
Ke' = K
e1 = q1
e2 = q2
r = r
Fe = Ke' q1 q2 / [r^2]
Hence "0" is the cause of the electrostatic force

Case3: Electromagnetic force:

The charged particle in a magnetic field,
v = w/r
if charged particle in magnetic field "B"
and v = 0 it can be held at that place
if v != 0 it can be held at r = w/v
r - r1 = 0
It has to be held by a centipetral force
F = k c1 c2 x ... x cn
causes are q, B and for a centipetral force v x sin x [theta]
F = q B v sin[theta]
v is the speed of the particle (magnitude of the velocity),
B is the magnitude of the magnetic field,
theta is the angle between the velocity and the magnetic field vectors.

qvB=mv^2/r
r1=[mv]/[qB]

Here, r is the radius of curvature of the path of a charged particle with mass m and charge q, moving at a speed v that is perpendicular to a magnetic field of strength B.

r - r1 = 0

"0" holds or is the cause of the sustaining force, sustaining here the charged particle in a circular motion of radius "r1"
Hence "0" is the cause of the electromagnetic force

The Yukawa potential for "strong force" and "weak force":

We consider properties of the Yukawa potential,

V (r) = [k/r] x e ^ [-r/alpha] = [k/r] x [1 / e ^ [r/alpha]]
Cn1 = 0 => r = 0 => 1/r = inf => F . s' => V ~ inf
Cn1 = inf => r = inf => 1/r = 0 => F . s' => F x s' x cos[theta] ~ F x s' => V~0
"0" or "1/r" and "e1" and "e2" are the cause of the yukawa potential
e1 and e2 are the entities between which the Yukawa potential holds
r is the distance between them
"0" is contained in e1 and e2

Conclusion:

The four forces are ultimately the force of attraction by 0's in like entities hence they are the same underlying force in essence.

TWELVE
SUMMARY

Law of action and reaction:

$E - E = 0$

E = Action

-E = Reaction

Action and Reaction are equal and opposite

Law of action and consequence:

$E\_t12 - E\_t34 = 0$

E_t12 = Action performed at time t12

E_t34 = Action experienced at time t34

What is committed is experienced

Law of cause and effect:

$E/E = [C1/C1] \times \ldots \times [C2/C2]$

E = Effect

C1 = cause1

C2 = cause2

Law of duality:

$E \times 1/E = 1$

E = entity E

1/E = opposite of entity E occur together

Action/Change Equation:

$[\, x + \{ [p1/ap1] \times [tw1] - [p2/ap2] \times [tw2] \} \,] - \{xw\} = \{0\}$

$p1/ap1 > p2/ap2$ Increase

$p1/ap1 < p2/ap2$ Decrease

$p1/ap1 = p2/ap2$ Being
$p1/ap1 < p2/ap2$ till 0 | Removal

pn = “acto” quality (controlled)
apn = “resisto” quality (controlled)
x = quality (controlled)
tw1 = controlled time
tw2 = controlled time
{0} = controller
{xw} = value willed by controller "0"

Illumination Equation:

$vl / vl = [L1 / L1] \times \ldots \times [ap1 / ap1]$

L1 = “Lumo” quality
ap1 = “resisto” quality

$\{ [L1/ap1] \times [tw1] \} - \{Iw\} = \{0\}$

{0} = controller
{Iw} = value willed by controller "0"

The cosmic form:

$O1 = O1x1 = q1A1$
$O2 = O2x1 = q2A2$
$On = Onx1 = qnAn$

$Ut2/Ut1 = O1/O1 \times O2/O2 \times \ldots \times On/On$
$Ut2 = [O1 \times 1] \times [O2 \times 1] \times \ldots \times [On \times 1]$
$1 \times 1 \times 1 \times .. \times 1 = 1$
$Ut2 = [O1] \times [O2] \times \ldots \times [On] \times [1]$
$Ut2 = [q1A1] \times [q2A2] \times \ldots \times [qnAn] \times [1]$

1 has all qualities
1 performs all actions
all are the qualities, actions and parts of "1"
O1 can be your sister and O2 can be your friend’s sister,
but they are one as it is only 1 acting through them
something which can be stated as "All Indians are my brothers and sisters"

The medicine law:

if an entity "B" causes disease, the same entity in a "non lethal form" or "dose" cures the disease
$C \sim 0 \mid 1/C \sim Inf$

C = disease causing ability
1/C = disease resisting or curing ability

The trade-economy equation:

{P – {Pw}} + [L][1]{ O‘ - O’ } = 0
P = person
Pw = Selected person
O‘ = {O1, O2, O3, O4 ....} for L = 1
O’ = {O1’, O2’, O3’, O4’ ....} for L = 0.5
O’ = Exchanged object or value of exchanged object depends on the value of L
L = 1 means Object O’ exchanged with 1
L = 1 means Object O’ exchanged with 0.5 “1”

3. **Law of cause and effect:**

E/E = [C1/C1] x ... x [Cn/Cn]
E = Effect
C1 = cause_1
Cn = cause_n

E - E = [C1 x C2 x .. x Cn] - [C1 x C2 x .. x Cn] = Ot+1@x - Ot@x = {0}
E = qg, qd, A, Obj
{0} is the controller or the cause of the cause and hence effect at a particular location x and time t
Obj = [qg qd A]

qg = C1 x ... x Cn / C2 x ... x C2n = C1n / C2n
qd = C3 x ... x Cn / C4 x ... x C4n = C3n / C4n

A = Modification and Movement
Del q = k x [C3 / C4] x t
Del xm = k x [C3 / C4] x t

www.ingramcontent.com/pod-product-compliance
Ingram Content Group UK Ltd.
Pitfield, Milton Keynes, MK11 3LW, UK
UKHW062312290726
14090UKWH00018B/1019